James Webb Space Telescope, a new Era in Astronomy

Patrick H. Stakem

Number 37 in Space Series

(c) 2021

Table of Contents

Table of Contents

Introduction

This is the story of the pending James Webb Space Telescope journey, written as its launch is near. It has been in work for quite a while. The project kicked off in 1996, some 25 years ago as of this writing. It is a NASA Project with the cooperation of ESA and the Canadian Space Agency. Fifteen countrys have contributed to the Project. The Project has four main goals. These are to search from the first stars and galaxies that formed after the Big Bang; to study the formation and evolution of galaxies, to contribute to the understanding of star and planetary system formation; and to study other planetary systems for signs of life.

Author

The Author received his first telescope while in grade school, and was hooked on exploring space from then on.

Mr. Patrick H. Stakem has been fascinated by the space program since the Vanguard launches in 1957. He received a Bachelors degree in Electrical Engineering from Carnegie-Mellon University, and Masters Degrees in Physics and Computer Science from the Johns Hopkins University. At Carnegie, he worked with a group of undergraduate students to re-assemble, modify, and operate a surplus missile guidance computer, which was later donated to the Smithsonian.

He began his career in Aerospace with Fairchild Industries on the ATS-6 (Applications Technology

Satellite-6) program, a communication satellite that developed much of the technology for the current TDRSS (Tracking and Data Relay Satellite System). He followed the ATS-6 Program through its operational phase, and worked on other projects at NASA's Goddard Space Flight Center including the Hubble Space Telescope, the International Ultraviolet Explorer (IUE), the Solar Maximum Mission (SMM), FUSE, some of the Landsat missions, and Shuttle. He was posted to NASA's Jet Propulsion Laboratory for Mars-Jupiter-Saturn (MJS-77), which later became the *Voyager* mission, and is still operating and returning data from outside the solar system at this writing. He initiated and lead the international Flight Linux Project for NASA's Earth Sciences Technology Office. He is the recipient of the Shuttle Program Manager's Commendation Award, and has completed 42 NASA Certification courses. He has two NASA Group Achievement Awards, and the Apollo-Soyuz Test Program Award.

Mr. Stakem has been affiliated with the Whiting School of Engineering of the Johns Hopkins University since 2007. He supported the Summer Engineering Bootcamp Projects at Goddard Space Flight Center for 2 years, and presented several cubesat summer courses.

I never worked on JWST, but saw it in the clean room numerous times, as I escorted student tours. I did work on the predecessor HST, Hubble Space Telescope and OAO, Orbiting Astronomical Observatory.

As I write this, the JWST spacecraft has arrived at the launch site in French Guiana. It came from a Northrop test site in Redondo Beach, California, through the Panama Canal, a journey of 5800 miles. It went to Naval Weapons Station, Seal Beach for loading on the RoRo Cargo ship MN Colibi. . It went through the Panama Canal, and south to Pariacaho, in French Guina. It traveled in a special container, STTARS, essentially a mobile cleanroom There is an airfield, but it is 40 miles from the launch site, and has 7 tight bridges. It is currently scheduled to launch on December 18, 2021 on an Ariane rocket. It is the latest in space-based telescopes, referred to as NGST – Next Generation Space Telescope.

Who was James E. Webb?

James Webb was NASA's second administrator, serving from 1961 to 1968. He was appointed by President Kennedy, and held the position during the Mercury and Gemini Programs, leaving just before the first Apollo flight. He had previously been Undersecretary of State, under Acheson. He was a Marine Corps pilot in the 1930's. He had worked for Sperry Gyroscope. In 1944, he re-enlised in the Marines. He was to leave for Japan in August of 1945, but his trip was canceled due to the end of the war. He had a law degree from George Washington University.

Mr. Webb passed away in 1992, not seeing his namesake launched.

Why Space Telescopes?

Earth's atmosphere limits and interferes with light coming through it. Our generated pollution is not helping any. We have seen about all we can see in our neighborhood (solar system) but we want to know more about other neighborhoods.

Previous Space Telescopes

This book does not specially discuss previous space telescope programs. My book by that title addresses 45 or those. I do discuss Hubble, since it seems to be doing well, and should be able to take simultaneous images with JWST.

Astronomy missions are usually designed to observe in specific spectral bands such as infrared or ultraviolet. Now we can see more clearly, and farther. The atmosphere not only causes scintillation, but filters higher frequencies such as the ultraviolet, x-rays, and gamma rays. It we want to see in these wavelengths, we need to get up above the atmosphere. Essentially, the atmosphere has "windows" in the optical and radio frequencies. Using high altitude aircraft and balloon-borne telescopes is only a partial solution.

Earth-based telescopes have the problem that they have to look through the atmosphere. One way to solve that is to put them in orbit. This was suggested in 1923 by Rocket Pioneer Hermann Oberth.

We have learned a lot about our solar system, galaxy, and the universe with telescopes in general.

Astronomy dates back far in the history of mankind, starting with observations of the sky, and the development of theories of what it all meant, and how it all worked. The initial assumption was that the Earth was the center of the Universe, and everything traveled around it. But, by observations, some "heavenly bodies" wander about in the sky, and increasingly complex theories were developed to address this. Later, when the Earth was dethroned as the center of the universe, it was understood that other planets and the Earth were in orbit around the Sun. And, the Sun was just another star, like those that could be seen in the night sky. After a while. Some "stars" were identified as planets, "wanderers" against the "fixed star" background. With more observations and better optics, we saw that other planets had moons as well. Not very long ago, we saw definitely that stars far away and no a part of out solar system, had planets as well, and they in turn had moons. What else are we going to find out?

ExoPlanets

Exoplanets are planets orbiting another star than our own Sun. Although it is difficult to see them through a telescope, we can define them by their gravitational effects on their primary (star). With JWST, we have expanded our search zone by a huge factor.

And there are exo-solar systems, with collections of

exoplanets, with their exo-moons. It is a strange Universe out there. We now see that planets are rather common in our galaxy. The next big question is, do any of them have life? It is postulated that there are some 11 billion exoplanets. We have only looked at a handful. As of this writing, there are more than 3,800 known exoplanets, with more being added every day. There are 600 known multi-planet systems.

Exoplanets are the current Astronomy hot topic, and dedicated missions have been launched to search for and characterize them. The number of interesting known exoplanets is changing on a daily basis. When the new James Webb Space Telescope gets launched, the floodgates will open wider.

Search for Exoplanets

One of the more exciting missions is the search for planets of other stars than our own sun. Although there are nearly impossible to image directly, they can be observed as the pass through our line of sight with the distant star, and cause a small dip in the perceived brightness. Exo-planets are best seen from space.

A few thousand have been cataloged. Follow-on questions are, do exo-planets have exo-moons, and can they harbor life? To have the potential of life, the planets have to orbit the right type of star, at the right distance, called the Goldilocks zone (from a Fairy tale). Not too hot, not too cold, just right.

Exoplanets are the current hot topic, and dedicated missions have been launched to search for and characterize them. JWST is the latest tool to explore and characterize ExoPlanets.

The Drake Equation

The Drake equation, formulated by Dr. Frank Drake in 1961, is a way of estimating the number of extraterrestrial Civilizations that can communicate. It was formulated by Dr. Drake at the Greenbank Radio observatory. The equation multiplies these factors:

$R*$, the average rate of star formation in our galaxy.

f_p, the fraction of those stars that have planets.

n_e, the average number of planets that can potentially support life per star that has planets.

f_l, the fraction of planets that could support life that actually   develop life at some point.

f_i, the fraction of planets with life that actually go on to develop   intelligent life.

f_c, the fraction of civilizations that develop a technology that   releases detectable signs of their existence into space.

L, the length of time for which such civilizations release detectable signals into space.

N , the number of civilizations in our Galaxy that ca
communicate.

In 1961, guessing at the parameters, the value for N was
10^3-10^8. As more data is collected, better values for the
parameters evolve. The current value is around 15
million.

The resultant figure for the universe suggests it is highly
unlikely that Earth hosts the only intelligent life that has
ever occurred. Maybe we can check with our neighbors...

The Fermi paradox, named after physicist Enrico Fermi,
is the apparent contradiction between the lack of
evidence and high probability estimates for the existence
of extraterrestrial civilization. What are we missing?

Initially, we had to guess at most of these parameters, but
better observation has brought about better results. As our
observation technology gets better, we get better numbers
for the factors in the equation.

In 1992, the first discovery that our solar system was not
unique in the Galaxy was made. This gave us some real
numbers to plug into the Drake equation. This
observation was done by radio-telescope.

The Red Dwarf star Trappist has a system of 7 planets,
nearly all Earth size. That a red dwarf would have a

planetary system was a surprise, but opened up the options for further research. At the moment, we can only detect exoplanets of Earth-size or larger.

Dr. Rene Doyon of the University of Montreal, with an International Team, photoed three XO planets orbiting the same star. This is the first detected X0 solar system. He used the Keck and Gemini North telescopes in Hawaii, on the summit of the volcano Mauna Kea. Both of these have 10 meter primary mirrors.

Alpha Centauri's planets

A special target for looking for exoplanets is Alpha Centauri, the Stat closest to out solar system, a mere 4.37 light years distant. It is actually a binary star system of Alpha Centauri A and Alpha Centauri B. There is an associated red dwarf, about 13,000 AU distant, called Alpha Centauri C, or Proxima Centauri. The European Southern Observatory discovered an Earth-sized planet orbiting Proxima Centauri in the habitable zone. This discovery came in 2016. Another small planet, smaller than Mercury, had been found in 2013, around Alpha Centauri. In 2015, the HST saw a transit event caused by a planet of about Earth's size. There may be more.

Since exoplanets do not emit light, but only reflect the light of their primary, it is hard to image them. It is easier in the infrared spectrum, but that radiation doesn't make it through Earth's atmosphere. That's where JWST comes in.

JWST

The James Webb Space Telescope is the follow-on to the Hubble Space Telescope. It uses updated technology and a new approach for the mirror, using 18 hexagonal segments, that are individually adjustable. There are 126 small motors to adjust the optics for fine tuning. Ball Aerospace was the prinicpal optical contractor. The resulting mirror is 6.5 meters in diameter. JWST observes in long wavelengths visible through the mid-infrared. The spacecraft will be placed in a halo orbit at the Earth-Sun L2 Lagrange point about 1.5 million miles from Earth. It has a large sun shield to block the Sun's light from interfering with the observations. The project was the top pick in NASA's 2000 Decadal Survey. Work has been going on since 1989, primarily at the Goddard Space Flight Center, the lead center for the project.

The spacecraft uses Spacewire internally for data communication. The onboard Solid state recorder has a capacity of 59 gigabytes.

The Integrated Science Instrument Module provides power, computing resources, and data storage. It is also a critical structural element. There are four science instruments, and a guide camera. The NIRCam instrument is an infrared imager working from the visible to the near infrared. It was built by the University of Arizona. The NIRSpec uses spectroscopy over the same range. It is a contribution from ESA, and was built in the Netherlands. The mechanisms and optical elements were provided by Carl Zeiss GmdH. The mid-infrared instrument observes in the 5 to 27 micrometer range. It has a camera and an imaging spectrometer. It was designed and built by a consortium of NASA and several European countrys. MIRI has to stay below 6 degrees K, and does this with a helium gas mechanical cooler. Both NIRCam and MIRI have blocking coronagraphs, that allow imaging of exo-planets, by blocking the light from their star.

The Fine Guidance Sensor, and Near Infrared Imager, also has a slitless spectograph. This was provided by the National Research Council of Canada. The Canadian Space Agency also provided a Near Infrared imager and slitless spectrograph for the 0.8 to 5 micrometer band.

JWST was not designed to be serviceable. It has a projected lifetime of 5 years, with a goal of twice that. Originally estimated to cost a half-billion dollars in 1997, the current cost looks in the vicinity of 8.8 billion dollars. Actually, the spacecraft will include a docking ring for an Orion crewed spacecraft (which has yet to fly), so in theory, it might be serviceable. It's predecessor, Hubble, if you recall, was serviced twice in orbit by the Shuttle. JWST uses small thrusters to maintain its orbit, and this might be its final downfall, if we can't replenish it.

One of the more interesting tasks for the observatory will be to assist in the search and characterization of exo-planets. It will be able characterize the atmospheres of potentially habitable exoplanets.

Besides the instruments, JWST has the standard engineering components for power, command and data handling, telemetry, and attitude control. The structure of the bus weighs more than 750 pounds, and supports the 13,700 pound telescope. This is called the spacecraft bus.

JWST will be operated from the same facility that Hubble is, the Space Telescope Science Institute, on the campus of the Johns Hopkins University in Baltimore, MD. This is also the Science and Operations Center for Hubble. Data to and from JWST will use NASA's Deep Space Network. Data will be processed and calibrated, and then distributed online to astronomers world wide. If you are a well known astronomer in the community, you can submit proposals for observations. Astronomers will have 1 year of exclusive access to the data, after which it will be put in a public online archive. There is a General Observers Program, a Guaranteed Time Observation Program, and A Director's Discretionary Early Release Program. Hubble will produce more than 450 Gigabytes of data per day.

There was a problem with the deployment test of the sunshield that took some time to resolve and correct. Nominal mission life is ten years. One limitation will be the station-keeping propellant supply. Another possible problem is a launch slip. Hopefully, HST will still be operational when JWST makes it to orbit, and the two can be used together. The ESA telescope Herschel will be in the vicinity of L2 when JWST arrives.

Why observe in the infrared, when it's hard? Mostly because space dust blocks visible light.

JWST is a joint project of NASA, ESA, and the Canadian Space Agency. In all, fourteen countries were involving in construction of the spacecraft. John Mather, the Senior Project Scientist, had previously received the 2006 Nobel Prize in Physics.

Partially to avoid the HST problem with the main mirror, the JWST has a large set of small, adjustable mirrors, individually controllable with 6 actuators each. They are gold-coated beryllium. Since it observes in the infrared part of the spectrum, the detectors have to be cooled to single digits above absolute zero. A large sun shield will be used to shadow the spacecraft form the Sun, Earth, and moon. The solar wind will push the sun shade and thus the spacecraft around. But, JWST has a trim tab to counter this.

The project was a concept in the mid-1990's and construction was completed in November of 2016. It is in testing at the moment. It was originally called the Next Generation Space Telescope (NGST), but was named after James Webb, the second NASA administrator.

Where Hubble has a 2.4 meter single-piece mirror, JWST has an 18 segment mirror for a combined size of 6.5 meters. These segments are folded for launch, and deploy in orbit. It has curved secondary and tertiary mirrors. Instruments include a near infrared camera, a near infrared spectrograph, a mid-infrared instrument, and a near infrared imager with a slit-less spectrograph. The telescope is expected to see extra-solar planets (planets around other stars). The telescope masses 6.5 tons. It will use NASA's Deep Space Network for data transmission.

The instrument set includes a near-infrared camera (NIRCam), and spectrograph (NIRSpec). These have 32 megapixels. It uses a new type of gyro, the hemispherical resonance gyro, for attitude sensing. With essentially no moving parts, this type of gyro should last a lot longer. The spacecraft has about 60 gigabytes of onboard storage, and uses lithium-ion batteries. Onboard, it runs Java scripts for operations.

It is not designed to be serviceable. It has a projected lifetime of 5 years, with a goal of twice that. Originally estimated to cost a half-billion dollars in 1997, the current cost looks in the vicinity of 8.8 billion dollars. Actually, the spacecraft will include a docking ring for an Orion manned spacecraft (which has yet to fly), so in theory, it might be serviceable.

One of the more interesting tasks for the observatory will be to assist in the search and characterization of exo-planets

Delivery of the spacecraft for Integration and Test at Northrop Grumman was delayed into 2019, with the launch slipping to 2021. There was a problem with the deployment test of the sunshield. Nominal mission life is ten years. One limitation will be the station-keeping propellant supply. Another possible problem is a launch slip. The 6200 kg spacecraft is scheduled to go on an Ariane-5 vehicle,

If HST is still operational when JWST makes it to orbit, the two can be used together. The ESA telescope

Herschel will be in the vicinity of L2 when JWST arrives. The launch was pushed back to Current launch date is At this writing, the launch is scheduled for Decemeber 18, 2021.

I was able to visit JWST several times, when it was in Goddard's massive clean room. Some of the components were tested in Goddard's Space Environmental Simulator (thermal-vacuum chamber). The entire telescope wouldn't fit, though. Later in the program, JWST was taken to the large chamber at Johnson Space Center, the largest in the world, developed for the Apollo program.

JWST will be placed in a position to shield it from Earth's Infrared emissions. It is going to the L2 Earth-Sun Lagrange Point. Let me explain this.

There is a point, on a line between the Earth and the Sun, where the gravity of the two body's cancel out. Strangely, a satellite can actually orbit that point, just as if there was a mass there. In the "Three body problem" of orbital mechanics, there are five such points. The one between two bodys is the easiest to grasp. That is not where JWST is going. It's a little more complicated than that.

In the larger sense, there is a point behind each primary (such as the Sun, moon, or a planet) and one off to each side. The points are nulls in the combined gravity field. (feel free to jump forward at any time). So, JWST is at the L2 point of the Sun-Earth system. This is located at 1.5 million miles further away from the Earth than the

Moon. In actuality, JWST shares Earth's orbit around the Sun, at a distance. It is currently impossible to service the spacecraft, so we have to get it right.

Space Weather

Space Weather impacts spacecraft. Since JWST is fairly fat from Earth, we don't have the problems of being close to an atmosphere. However, being far from Earth removes some of the natural shielding.

The Sun controls the weather in our solar system. Large solar flares can release up to 10^{25} joules of Energy. The Sun releases electrons, stripped from their atoms, the resulting ions, and intact atoms from the corona. It also releases radio waves, which, traveling at the speed of light, reach Earth 8 minutes later. The particles travel at sub-light speeds. Bright aurora's in the polar regions will be created. Stellar flares have also been observed on other stars.

When the solar wind reaches the Earth's magnetic field, it interacts with it, creating a Geomagnetic storm. There can also be proton storms from the Sun. One effect is that the upper atmosphere is heated to tens of millions of degrees Kelvin. Luckily, it is near-vacuum. Increased electromagnetic radiation from radio to Gamma Rays are also observed.

There are sentinal satellites continuously monitoring the Sun, and its impact, the solar wind. We can get a 2-hour heads-up on charged particles coming our way.

The bow shock is caused by plasma from the Sun hitting

the Earth's magnetosphere. The same effects are observed at other planets with magnetic fields, Jupiter, particularly. The plasma is ionized, and follows spiral paths along the planet's magnetic field lines. The flow speed, at Earth, is around 400 km/s. The shock, at Earth, is some 17 km thick, and located 90,000 km Sun-ward. Bow shocks have been observed on planets in other star systems.

One issue with spacecraft being enveloped in a big burst of charged particles is charging. This can be a problem when an electrical potential develops across the structure. Ideally, steps were taken to keep every surface linked, electrically. But, the changing phenomena has been the cause of spacecraft system failures. Where does the charge come from? Mostly, the Sun, in the forms of charged particles. This can cause surface charging, and even internal charging. Above about 90 kilometers in altitude, the spacecraft is in a plasma environment At low Earth orbit, there is a low energy but high density of the plasma. The plasma rotates with the Earth's magnetic field. The density is greater at the equator, and less at the magnetic poles. Generally, electrons with energies from 1-100 keV cause surface charging, and those over 100 keV can penetrate and cause internal charging. As modern electronics is very susceptible to electron damage, proper management of charging is needed at the design level.

Just flying along in orbit causes an electric field around the spacecraft, as any conductor traveling through a magnetic field does. If everything is at the same potential, we're good, but if there's a difference in potential, there can be electrostatic discharge. These discharges lead to electronics damage and failure, and can also cause physical damage to surfaces, due to arcing. This has been a problem at the International Space Station.

During periods of intense solar activity, Coronal Mass Ejection (CME) events can send massive streams of charged particles outward. These hit the Earth's magnetic field and create a bow wave. The Aurora Borealis or Northern Lights and the Aurora Australias or Southern Lights are manifestations of incoming charged particles hitting the upper reaches of the ionosphere.

Galactic Cosmic rays are actually heavy ions, originating outside of our solar system. The actual origin is unknown. They carry massive amounts of energy, up into the billions (10^9) of electron volts.

Cosmic rays, particles and electromagnetic radiation, are omni-directional, and come from extra-solar sources. Some 85%, are protons, with gamma rays and x-rays thrown in the mix. Energy levels range to 10^6 to 10^8 electron volts (eV). These are mostly filtered out by Earth's atmosphere. There is no such mechanism on the Moon, where they reach and interact with the surface material. Solar flux energy's range to several Billion electron volts (Gev).

JWST, out at its Libration point, has no magnetic field to

redirect charged particles. It is essentially in the same orbit as the Earth, and can be shut down if a massive storm is coming out from the Sun. That would be done to protect the sensitive electronics. The spacecraft has been designed with proper grounding, which is to say, are surfaces are connected, and should be at the same potential.

And, other approaches

Another approach to avoid observing through Earth's atmosphere is an observatory on the Moon. The International Lunar Observatory is a commercial facility to be placed at the lunar south pole, possibly as soon as 2023. It was originally planned for 2008. Another projected idea is a radio telescope at the lunar pole, at least, the lunar backside. On the Moon you get 14 days of observation, followed by 14 days to analyze your data.

Another planned Lunar telescope is the Lunar Ultraviolet Cosmic Imager (LUCI), a project of the Indian Institute of Astrophysics. It will be deployed by a commercial lunar lander. The original launch data was 2020, but the launch services company dropped out.

If a new launch provider steps forward, LUCI could provide valuable information of Near-Earth Objects, and potentially hazardous objects. UV radiation does not penetrate Earth's atmosphere, so an Earth-orbiting or lunar mission is required. The telescope instrument has been ready since 2019.

Afterword

Shortly, it is going to get very exciting in the field of Astronomy, as we get more and better data about other stars and their planets and moons.

Glossary of Terms

See also, https://en.wikipedia.org/wiki/List_of_astronomy_acronyms

AAS – American Astronomical Society.

ABRIXAS - Broadband Imaging X-ray All-Sky Survey.

AGILE-Astro-Rivelatore Gamma a Immagini Leggero (Italy)

AGN – active galactic nucleus with a relativistic jet of ionized   matter.

Alexis - Array of Low Energy X-ray Imaging Sensor.

Amor asteroids – a class of near Eartha asteroids.

AMS - Alpha Magnetic Spectrometer.

Angstrom – measure of length, 0.1 nanometer.

ANS - Astronomical Netherlands Satellite.

AO – adaptive optics

APL – Applied Physics Lab of the Johns Hopkins University.

Apogee – furthest point from a primary, in an orbit.

Apollo asteroids – class of near-Earth crossing asteroids

Apsys – extreme point in an orbit, closest or furthest.

Arc-minute – 1/60 of a degree.

Arc-second – 1/3660 of a degree.

ASCA - Advanced Satellite for Cosmology and Astrophysics, Japan.

ASIC – Application specific integrated circuit.

ASC- LPI - Astro Space Center of PN Lebedev Physics

Institute, Russian Academy of Sciences.

ASIN – Amazon Standard Inventory Number.

ASIS - Advanced CCD Imaging Spectrometer (Chandra).

ASM – all sky monitor

Asteroid – a chunk of rock; a minor planet.

Asteroid belt – a disc in the solar system between Mars and Jupiter, occupied by chunks of rock in various sizes.

Astrometry – measurements of positions and movements of objects in space.

Asteroseismology – study of oscillations in stars.

Aten asteroid – Earth crossing asteroids.

Atira asteroids – orbits entirely within Earth's orbit

AU – astronomical unit, mean distance from the Earth to the Sun, 93,000,000 miles.

AXAF - Advanced X-ray Astrophysics Facility (Chandra).

BATSE - Burst and Transient Source Experiment, (CGRO).

BBXRT – Broad Band X-ray telescope.

Big Bang – current cosmological model for the Universe.

Binary star – two stars in orbit around a common point.

Black hole – a collapsed star, compressed so dense that not even light can escape; a singularity.

Blazar – active galactic nucleus with a relativistic jet

Blue Moon – intercalary moon, 13 full moons in a year instead of 12.

Blue Shift – an apparent shift of electromagnetic radiation toward decreasing wavelength.

BRITE – (U.K.) Bright Target Explorer

CCD – Charge Coupled Device (like in your cell phone

camera)

C&DH – command and data handling.

CDR – critical design review

Centaur – a minor planet in an unstable orbit, behaving like an asteroid or comet.

Cepheid – a variable star.

CERN - Conseil européen pour la recherche nucléaire , European Organization for Nuclear Research.

CGRO – Compton Gamma Ray Observatory

Cheops – ESA, Characterizing ExOPlanets Satellite

Cluster – groups of stars.

CMB – Cosmic microwave background.

CME – Coronal Mass Ejection – burst of plasma and magnetic fields from a star's corona.

CNES - Centre national d'études spatiales. (France)

CNRS – Centre National de la Recherche Scientifique (France)

CNSA – China National Space Administration.

COBE - Cosmic Background Explorer.

Coma – Comet's tail

Comet – a solar system object consisting of ice, dust, and gas, in highly eccentric orbit.

Constellation – patterns we see in collections of stars.

COROT (French) Convection, Rotation et Transits planétaires.

Cosmic ray – high energy radiation, from outside the solar system.

COSPAR – Committee on Space Research, International Council for Science.

CSA – Canadian Space Agency.

CTP – Command and Telemetry Processor

CXE – Cosmic X-ray experiment, HEAO-1.

CXO – Compton X-ray Observatory.

Cyrogenic – at very low temperature.

Dark energy – hypothetical form of energy that explains why the Universe is expanding.

Dark Matter – existence postulated. Might account for 85% of the matter in the known universe.

DARPA – (U.S.) Defense Advanced Research Project Agency.

Deuterium – isotope of hydrogen

DLR – (German) Deutsches Zentrum für Luft- und Raumfahrt.

DMC – data management center.

DOE – (U.S.) Department of Energy.

DTA – Deployable tower assembly.

Dwarf planet – planet below a certain size.

Dwarf star – small star, much smaller than our Sun. Comes in   white, red, blue and black variations.

EBL – Extragalactic background light.

Ecliptic – the apparent path that the Sun seems to follow, the same as the Earth's orbit.

EDU – engineering design unit.

EGRET - Energetic Gamma Ray Experiment Telescope (CGRO).

ELDO – European Launcher Development Organization, merged into ESA.

ELT – extremely large telescope.

EMR – electromagnetic radiation.

Equinox – 2 days per year when there are equal periods of daylight   and darkness.

ESA – European Space Agency.

ESRO – European Space Research Organization, merged into   ESA.

EU – European Union.

EUV – extreme ultraviolet, wavelengths from 10nm to 124nm.

EUVE - Extreme Ultraviolet Explorer.

ev – electron volt, unit of energy.

Exa- 10^{19}

Exeroid – proposed name for a asteroid-like body from out of outer solar system.

Exoplanet – planet in another solar system.

FGST – Fermi Gamma ray Space Telescope.

FSM – fine steering mirror (JWST)

FUSE - Far Ultraviolet Spectroscopic Explorer.

Galaxy – a loosely coupled collection of 10^8 to 10^{14} stars.

Galex - Galaxy Evolution Explorer.

Gamma ray – EMR from radioactive decay.

Gas giant – a large planet consisting mostly of hydrogen and   helium. Jupiter and Saturn.

GEV – giga (10^9) electron volts

GLAST - Gamma Ray Large Area Space Telescope.

GN – NASA's ground network.

GO – general observers

GOALS - Great Observatories All-sky LIRG Survey, (Spitzer, Chandra, CHIPS, HST)

Gravitational Wave – a disturbance in the fabric of space-time.

GRB – gamma ray burst

GRO – Gamma Ray Observatory.

GSFC – Goddard Space Flight Center.

GTC - Gran Telescopio Canarias

GTO – Guaranteed Time Observations.

HabEx - Habitable Exoplanet Imaging Mission

Hard x-ray – energy above 5-10 kev.

HEASARC - High Energy Astrophysics Science Archive Research Center.

HEAO – High Energy Astronomy Observatory.

Helioseismology – study of oscillations in our Sun.

Heliosphere – a volume of space dominated by the Sun, In our case, out beyond Pluto.

HEFT – High Energy Focusing Telescope.

Heliocentric – sun-centered.

Hexte – instrument on RXTE spacecraft – High energy x-ray timing experiment.

HRMA – High resolution Mirror Assembly (Chandra)

HST – Hubble Space Telescope.

Hubble – Space Telescope named after Edwin Hubble.

Hubble Constant – rate of expansion of the universe.

HUDF – Hubble Ultra-Deep Field.

HUT - Hopkins Ultraviolet Telescope

HXMT – (China) Hard X-ray Modulation Telescope.

IAU – International Astronomical Union.

IBEX - Interstellar Boundary Explorer.

ICDH – Instrument Command and Data Handling.

Ice giant – large planet consisting of ices of various substances – Uranus and Neptune.

ICRP – Independent Comprehensive Review Panel

ICSU – International Council for Science.

IIA – Indian Institute of Astrophysics.

ILO – International Lunar Observatory.

Infrared cirrus -cloud-like structures in space that emit infrared light.

INTA - Instituto Nacional de Técnica Aeroespacial, Spain.

Integral - INTErnational Gamma Ray Astrophysics Laboratory

IPAC – (NASA's) Infrared Processing and Analysis Center.

IPN – Interplanetary Network – spacecraft with gamma ray burst detectors.

IRAC - Infrared Array Camera (Spitzer)

IRAS – Infrared Astronomical Satellite.

ISIM – Integrated Science Instrument Model.

IRIS - Interface Region Imaging Spectrograph.

IRS - Infrared Spectrograph (Spitzer).

IRSA - Infrared science archive.

IRT – Infrared Telescope (Shuttle mission).

ISAS – (Japan) Institute of Space and Astronautical Science.

ISBN – International Standard Book Number.

ISIM – (JWST) Integrated Science Instrument Module

ISM – interstellar medium

ISO – Infrared Space Observatory; International Standards Organization.

ISRO – Indian Space Research Organization.

ISS – International Space Station.

IXAE – (India) X-ray Astronomy Experiment.

IXPE - Imaging X-Ray Polarimetry Explorer.

JAXA – Japanese Aerospace Exploration Agency.

JDEM – Joint Dark Energy Mission (NASA, DOE)

JHU – Johns Hopkins University.

Jovian – pertaining to Jupiter.

JPL – NASA's Jet Propulsion Lab, Pasadena, California.

JWST – James Webb Space Telescope, also Just Wonderful Space Telescope.

Kaistsat - Korea Advanced Institute of Science and Technology Satellite.

KAO – (NASA) Kuiper Airborne Observatory.

KARI – Korea Aerospace Research Institute.

Kev – kilo-electron volts, a measure of energy

KBO – Kuiper Belt objects.

KOI – Kepler Objects of Interest.

Kuiper Belt – disc in the solar system from Neptune out about 50AU.

Lagrange point – null in the gravity field.

LANL – (U. S.) Los Alamos National Laboratory.

LASP - Laboratory for Atmospheric and Space Physics (U. Colorado)

LASS - Large-Area Sky Survey, HEAO-1.

LEGRI - Low Energy Gamma-Ray Imager.

Leica – low energy ion composition analyzer

LEO – low Earth orbit.

LETGS - Low Energy Transmission Grating Spectrometer (Chandra)

LGM – little green men.

Light pollution – interference from background sources.

Light year – the distance light travels in one year. 9.5 $x10^{12}$ kilometers.

LIRG – luminous infrared galaxy.

LISA – Laser Interferometer Space Antenna.

LMSAL - Lockheed Martin Solar and Astrophysics Laboratory.

LRR - launch readiness review.

LUCI – Lunar Ultraviolet Cosmic Imager (India)

Lunation – synodic month; average period of the moon's rotation.

LUT – (China) Lunar-based ultraviolet telescope.

LUVOIR - Large UV Optical Infrared Surveyor.

LWS – Living with a Star, NASA Program.

Magellanic clouds – two dwarf galaxies in the southern sky.

Magnetar – neutron star with powerful magnetic field

Magnetotail – a long stream of charged particles, held by magnetic   forces.

MAST - Mikulski Archive for Space Telescopes (named for a Maryland Senator)

Mbps – 10^6 bits per second.

Mbytes – mega (10^6) bytes.

MCC – Mission Control Center

MCDR – Mission Critical Design Review

MEV – million electron volts, a measure of energy

microlensing – light being bent by the gravity of a large body.

micron – micro meter

Milky Way – our Sun is in this Galaxy.

MIPS - Multiband Imaging Photometer (Spitzer)

MIT – Massachusetts Institute of Technology.

MMOC – Multi-Mission Operations Center, GSFC, Bldg 14.

MMS – multimission modular spacecraft

Moon – smaller astronomical body in orbit around a planet.

MSFC - Marshall Space Flight Center.

Nanometer – 10^{-9}

NASCOM – NASA communications network, basement,

Bldg 14.

NASDA – National Space Development Agency (Japan)

NEA – near Earth asteroid.

NEC – near Earth comet.

NASA – National Aeronautics and Space Administration.

Nebula – interstellar cloud of dust and gasses.

NeN – (NASA) near Earth network

NEO – near Earth object.

Neutron star – collapsed core of a large star. Very dense.

New Horizons – imaging mission to Pluto and beyond.

NGC – New General Catalog (of Nebulae and Clusters of Stars).

NIAC - NASA Institute for Advanced Concepts.

NICER – Neutron star interior composition explorer.

NICMOS – (HST) Near Infrared Camera and Multi-Object Spectrometer.

NM – nano-meter

Nova – transient astronomical event involving a bright new star that fades over time.

NRC – (Canada) National Research Council.

NRL – Naval Research Laboratory, Washington, D.C.

NSSDC -(U.S.) National Space Science Data Center.

NSF – (U.S.) National Science Foundation.

NSSC – NASA Standard Spacecraft Computer, an 18-bit flight computer.

NuSTAR - Nuclear Spectroscopic Telescope Array

OAO – Orbiting Astronomical Observatory.

Orbit – the path of one body around another, that are linked by gravity.

OSSE - Oriented Scintillation Spectrometer Experiment (CGRO).

OTE – optical telescope element

PAMELA - Payload for Antimatter Matter Exploration and Light-   nuclei Astrophysics.

Parsec – parallel second of arc, unit of length, about 3.26 light years.

PCA – proportional counter array.

PDR – preliminary design review.

Pera – nearest point to the primary, in an orbit.

pet – proton-electron telescope

Peta – 10^{16}

PHA – potentially hazardous asteroids

PHO – potentially hazardous objects

Photon – quantum particle of the electromagnetic field , zero mass,   moves at speed of light.

Planet – a body orbiting a star.

Planetary disk – debris disks around a star.

Pleides – an open star clusterPHA

Pulsar – highly magnetized rotating neutron star of a white dwarf.

Quark – an elementary particle.

Quasar – quasi-stellar object, galactic nucleus.

RAE - Radio astronomy explorer.

RC – High resolution camera (Chandra).

Red Shift – an apparent shift of electromagnetic radiation toward an increasing wavelength due to the doppler effect.

Ring system – a disk of solid material around a planet.

Rhessi - Ramaty High-Energy Solar Spectroscopic Imager.

Rockoon – a rocket taken to altitude by a balloon and then   launched.

Rogue planet – planet not associated with a star.

RoRo – roll on, roll off cargo ship.

ROSAT (short for Röntgensatellit), X-ray observer, German.

RTG – radioisotope thermoelectric generator.

RXTE – Rossi X-ray Timing Explorer.

Sampex - Solar Anomalous and Magnetospheric Particle Explorer.

SAO - Smithsonian Astrophysical Observatory.

SAS - Small Astronomy Satellite. (NASA)

Scintillation – variations in apparent brightness.

SD – scattered disk, a distant ring of smaller solar system objects, beyond Neptune

SDO – scattered disk object; Solar dynamics observatory

SERC – Science and Engineering Research Council (U.K.)

SES (Nasa-GSFC) space environment simulator.

SETI – search for extra-terrestial intelligence.

Seyfert galaxy – galaxy with quasar-like nuclei, about 10% of all   known galaxies.

SGR – soft gamma ray repeater; emits large bursts of gamma-rays and X-rays at irregular intervals.

SIM – Science Instrument Module; space interferometry mission

SIRTF - Space Infrared Telescope Facility, renamed Spitzer.

S&OC Science and Operations Center.

SMEX – Small Explorer program (NASA)

SMM – Solar Maximum Mission.

SN – (NASA) space network.

SNSA - Swedish National Space Agency.

SOFIA – NASA Stratosphere Observatory for Infrared Astronomy.

Soft x-ray – energies below 5 kev

Solar flare – a sudden rapid emission of electrons, ions, and atoms from a star.

Solar System – A star and its associated planets and such.

Solar wind – stream of charged particles emitted from a star's upper atmosphere.

Solo – (ESA) solar orbiter

Solstice – day of the shortest or longest period of daylight.

Spacewire – a communications protocol for networking on spacecraft.

SRC - Science Research Council (UK)

SRON – Netherlands Institute for Space Research.

SSC – Swedish Space Corporation.

SSR – solid state recorder.

STOL – system test oriented (computer) language.

Strangelet – a hypothetical particle, made of up, down, and strange quarks.

STS – Space Transportation System (Shuttle).

STTARS – Space Telescope Transporter for Air, Road, and Sea.

StScI – Space Telescope Science Institute (JHU)

Supernova – a massive explosion of a star, at its end of life.

SWG – Science Working Group.

SWOOPS -Solar Wind Observations Over the Poles of the Sun.

TBD – to be determined.

TDRS – Tracking and Data Relay Satellite.

TDRSS – Tracking and Data Relay Satellite System.
TESS - Transiting Exoplanet Survey Satellite.
Tidal lock – where the same side of a object always faces
    the primary it is orbiting.
TLP -transient lunar phenomena.
TNO – Trans-Neptunian objects.
Trojan – minor planet that shares an orbit with one of the
    larger planets.
TRW - Thompson Ramo Wooldridge, Aerospace Co.
UK – United Kingdom, England.
USAF – United States Air Force.
USRA – Universities Space Research Association.
UV – ultraviolet, 19 nm to 400 nm wavelength.
WFIrST - Wide Field Infrared Survey Telescope.
VIM – Voyager Interstellar Mission.
Virtual telescope - several robotic telescopes, remotely
    available in real-time over the Internet.
Wfirst - Wide Field Infrared Survey Telescope
White dwarf – very dense remnant of a stellar core.
WISE - Wide Field Infrared Survey Explorer.
WMAP - Wilkinson Microwave Anisotropy Probe.
Wmops - WISE Moving Object Processing Software.
WUPPE - Wisconsin Ultraviolet Photo-Polarimeter
    Experiment.
XMM - X-ray Multi-Mirror Mission.
X-ray - 0.1 to 10 nanometer wavelength.
X-ray binary (star) – binary star, emitting x-rays.
XRCF – X-ray & cyrogenic facility (MSFC)
YSO – young stellar objects.
Zombie-sat – dead satellite, in orbit.

Bibliography

Associated Press, *The Hubble Space Telescope: A Universe of New Discovery*, 2015, ISBN-1633530469.

Baker, David NASA *Hubble Space Telescope - 1990 onwards (including all upgrades): An insight into the history, development, collaboration, construction and role of ... space telescope* (Owners' Workshop Manual), 2015, ISBN-100857337971.

Copernicus, Nicolaus *On the Revolutions of the Heavenly Spheres,*
1543, translated from Latin to English, ASIN-B01MS8TGOV.

Devorkin, David H.; Smith, Robert *Hubble: Imaging Space and Time*, 2008,National Geographic, ISBN-1426203225.

Devorkin, David H. *The Hubble Cosmos: 25 Years of New Vistas in Space*, 2015, National Geographic, ISBN-9781426215575.

Dickinson, Terence *Hubble's Universe: Greatest Discoveries and Latest Images*, 2017, ISBN-9781770859975.

Ellerbroek, Lucas*, Planet Hunters: The search for extraterrestrial life,* 2017, ASIN-B073S986GV.

English, Neil *Space Telescopes: Capturing the Rays of the Electromagnetic Spectrum,* 2017, ISBN-978-3319278124.

Greene, Thomas P. James *Webb Space Telescope,* 2013, NASA Technical Reports Server, ISBN-978-1289058685.

Langley, Andrew, *Planet Hunting: Racking Up Data and Looking for Life,* 2019, ISBN-978-1543572704.

NASA, *NASA's Great Observatories,* 2013, ASIN-B00DJUALH0.

NASA, "NASA Astrophysics Missions: Reviews of Operating Missions - Hubble Space Telescope, Chandra X-ray Observatory, Fermi Gamma-ray Telescope, Kepler, Planck, Suzaku, Swift, Warm Spitzer, XMM-Newton." 2012, ASIN-B007SH9FCM.

Summers, Michael *Exoplanets: Diamond Worlds, Super Earths, Pulsar Planets, and the New Search for Life beyond Our Solar System,* 2017, Smithsonian Books, ASIN-B01HA426MS.

Resources

- Webb site: http://webbtelescope.org/

- "Complete Guide to NASA's James Webb Space Telescope (JWST) Project - Spacecraft, Instruments and Mirror, Science,

Infrared Astronomy, GAO and Independent Review Reports, Congressional Hearings, 2011, ISBN-1549878212.

- https://www.jwst.nasa.gov/content/webbLaunch/index.html

- https://spacese.spacegrant.org/JWST_Mission_Operations_Concept_Document.pdf

- wikipedia, various.

If you enjoyed this book, you might also be interested in some of these.

Stakem, Patrick H. *16-bit Microprocessors, History and Architecture*, 2013 PRRB Publishing, ISBN-1520210922.

Stakem, Patrick H. *4- and 8-bit Microprocessors, Architecture and History*, 2013, PRRB Publishing, ISBN-152021572X,

Stakem, Patrick H. *Apollo's Computers,* 2014, PRRB Publishing, ISBN-1520215800.

Stakem, Patrick H. *The Architecture and Applications of the ARM Microprocessors,* 2013, PRRB Publishing, ISBN-1520215843.

Stakem, Patrick H. *Earth Rovers: for Exploration and Environmental Monitoring,* 2014, PRRB Publishing, ISBN-152021586X.

Stakem, Patrick H. *Embedded Computer Systems, Volume 1, Introduction and Architecture*, 2013, PRRB Publishing, ISBN-1520215959.

Stakem, Patrick H. *The History of Spacecraft Computers from the V-2 to the Space Station*, 2013, PRRB Publishing, ISBN-1520216181.

Stakem, Patrick H. *Floating Point Computation*, 2013, PRRB Publishing, ISBN-152021619X.

Stakem, Patrick H. *Architecture of Massively Parallel Microprocessor Systems*, 2011, PRRB Publishing, ISBN-1520250061.

Stakem, Patrick H. *Multicore Computer Architecture,* 2014, PRRB Publishing, ISBN-1520241372.

Stakem, Patrick H. *Personal Robots*, 2014, PRRB Publishing, ISBN-1520216254.

Stakem, Patrick H. *RISC Microprocessors, History and Overview,* 2013, PRRB Publishing, ISBN-1520216289.

Stakem, Patrick H. *Robots and Telerobots in Space Application*s, 2011, PRRB Publishing, ISBN-1520210361.

Stakem, Patrick H. *The Saturn Rocket and the Pegasus Missions, 1965,* 2013, PRRB Publishing, ISBN-1520209916.

Stakem, Patrick H. *Visiting the NASA Centers, and Locations of Historic Rockets & Spacecraft,* 2017, PRRB Publishing, ISBN-1549651205.

Stakem, Patrick H. *Microprocessors in Space*, 2011, PRRB Publishing, ISBN-1520216343.

Stakem, Patrick H. Computer *Virtualization and the Cloud*, 2013, PRRB Publishing, ISBN-152021636X.

Stakem, Patrick H. *What's the Worst That Could Happen? Bad Assumptions, Ignorance, Failures and Screw-ups in Engineering Projects, 2014,* PRRB Publishing, ISBN-1520207166.

Stakem, Patrick H. *Computer Architecture & Programming of the Intel x86 Family, 2013,* PRRB Publishing, ISBN-1520263724.

Stakem, Patrick H. *The Hardware and Software Architecture of the Transputer,* 2011,PRRB Publishing, ISBN-152020681X.

Stakem, Patrick H. *Mainframes, Computing on Big Iron,* 2015, PRRB Publishing, ISBN- 1520216459.

Stakem, Patrick H. *Spacecraft Control Centers,* 2015, PRRB Publishing, ISBN-1520200617.

Stakem, Patrick H. *Embedded in Space,* 2015, PRRB Publishing, ISBN-1520215916.

Stakem, Patrick H. *A Practitioner's Guide to RISC Microprocessor Architecture,* Wiley-Interscience, 1996, ISBN-0471130184.

Stakem, Patrick H. *Cubesat Engineering,* PRRB Publishing, 2017, ISBN-1520754019.

Stakem, Patrick H. *Cubesat Operations*, PRRB Publishing, 2017, ISBN-152076717X.

Stakem, Patrick H. *Interplanetary Cubesats*, PRRB Publishing, 2017, ISBN-1520766173 .

Stakem, Patrick H. Cubesat Constellations, Clusters, and Swarms, Stakem, PRRB Publishing, 2017, ISBN-1520767544.

Stakem, Patrick H. *Graphics Processing Units, an overview*, 2017, PRRB Publishing, ISBN-1520879695.

Stakem, Patrick H. *Intel Embedded and the Arduino-101, 2017,* PRRB Publishing, ISBN-1520879296.

Stakem, Patrick H. *Orbital Debris, the problem and the mitigation*, 2018, PRRB Publishing, ISBN-*1980466483*.

Stakem, Patrick H. *Manufacturing in Space*, 2018, PRRB Publishing, ISBN-1977076041.

Stakem, Patrick H. *NASA's Ships and Planes*, 2018, PRRB Publishing, ISBN-1977076823.

Stakem, Patrick H. *Space Tourism*, 2018, PRRB Publishing, ISBN-1977073506.

Stakem, Patrick H. *STEM – Data Storage and Communications*, 2018, PRRB Publishing, ISBN-

1977073115.

Stakem, Patrick H. *In-Space Robotic Repair and Servicing*, 2018, PRRB Publishing, ISBN-1980478236.

Stakem, Patrick H. *Introducing Weather in the pre-K to 12 Curricula, A Resource Guide for Educators*, 2017, PRRB Publishing, ISBN-1980638241.

Stakem, Patrick H. *Introducing Astronomy in the pre-K to 12 Curricula, A Resource Guide for Educators*, 2017, PRRB Publishing, ISBN-198104065X.
Also available in a Brazilian Portuguese edition, ISBN-1983106127.

Stakem, Patrick H. *Deep Space Gateways, the Moon and Beyond*, 2017, PRRB Publishing, ISBN-1973465701.

Stakem, Patrick H. *Exploration of the Gas Giants, Space Missions to Jupiter, Saturn, Uranus, and Neptune*, PRRB Publishing, 2018, ISBN-9781717814500.

Stakem, Patrick H. *Crewed Spacecraft*, 2017, PRRB Publishing, ISBN-1549992406.

Stakem, Patrick H. *Rocketplanes to Space*, 2017, PRRB Publishing, ISBN-1549992589.

Stakem, Patrick H. *Crewed Space Stations,* 2017, PRRB Publishing, ISBN-1549992228.

Stakem, Patrick H. *Enviro-bots for STEM: Using Robotics in the pre-K to 12 Curricula, A Resource Guide for Educators,* 2017, PRRB Publishing, ISBN-1549656619.

Stakem, Patrick H. *STEM-Sat, Using Cubesats in the pre-K to 12 Curricula, A Resource Guide for Educators*, 2017, ISBN-1549656376.

Stakem, Patrick H. *Lunar Orbital Platform-Gateway*, 2018, PRRB Publishing, ISBN-1980498628.

Stakem, Patrick H. *Embedded GPU's*, 2018, PRRB Publishing, ISBN- 1980476497.

Stakem, Patrick H. *Mobile Cloud Robotics*, 2018, PRRB Publishing, ISBN- 1980488088.

Stakem, Patrick H. *Extreme Environment Embedded Systems,* 2017, PRRB Publishing, ISBN-1520215967.

Stakem, Patrick H. *What's the Worst, Volume-2*, 2018, ISBN-1981005579.

Stakem, Patrick H., *Spaceports*, 2018, ISBN-1981022287.

Stakem, Patrick H., *Space Launch Vehicles*, 2018, ISBN-1983071773.

Stakem, Patrick H. *Mars*, 2018, ISBN-1983116902.

Stakem, Patrick H. *X-86, 40th Anniversary ed*, 2018, ISBN-1983189405.

Stakem, Patrick H. *Lunar Orbital Platform-Gateway*, 2018, PRRB Publishing, ISBN-1980498628.

Stakem, Patrick H. *Space Weather*, 2018, ISBN-1723904023.

Stakem, Patrick H. *STEM-Engineering Process*, 2017, ISBN-1983196517.

Stakem, Patrick H. *Space Telescopes*, 2018, PRRB Publishing, ISBN-1728728568.

Stakem, Patrick H. *Exoplanets*, 2018, PRRB Publishing, ISBN-9781731385055.

Stakem, Patrick H. *Planetary Defense*, 2018, PRRB Publishing, ISBN-9781731001207.

Patrick H. Stakem *Exploration of the Asteroid Belt*, 2018, PRRB Publishing, ISBN-1731049846.

Patrick H. Stakem *Terraforming*, 2018, PRRB Publishing, ISBN-1790308100.

Patrick H. Stakem, *Martian Railroad*, 2019, PRRB Publishing, ISBN-1794488243.

Patrick H. Stakem, *Exoplanets,* 2019, PRRB Publishing, ISBN-1731385056.

Patrick H. Stakem, *Exploiting the Moon,* 2019, PRRB Publishing, ISBN-1091057850.

Patrick H. Stakem, *RISC-V, an Open Source Solution for Space Flight Computers,* 2019, PRRB Publishing, ISBN-1796434388.

Patrick H. Stakem, *Arm in Space*, 2019, PRRB Publishing, ISBN-9781099789137.

Patrick H. Stakem, *Extraterrestrial Life*, 2019, PRRB Publishing, ISBN-978-1072072188.

Patrick H. Stakem, *Space Command*, 2019, PRRB Publishing, ISBN-978-1693005398.

Powerships, Powerbarges, Floating Wind Farms: electricity when and where you need it, 2021, PRRB Publishing, ISBN-979-8716199477.

Hospital Ships, Trains, and Aircraft, 2020, PRRB Publishing, ISBN-979-8642944349.

2020/2021 Releases

CubeRovers, a Synergy of Technologys, 2020, ISBN-979-8651773138

Exploration of Lunar & Martian Lava Tubes by Cube-X, ISBN-979-8621435325.

Robotic Exploration of the Icy moons of the Gas Giants, ISBN- 979-8621431006.

History & Future of Cubesats, ISBN-978-1986536356.

Robotic Exploration of the Icy Moons of the Ice Giants, by Swarms of Cubesats, ISBN-979-8621431006.

Swarm Robotics, ISBN-979-8534505948.

Introduction to Electric Power Systems, ISBN-979-8519208727.

Centros de Control: Operaciones en Satélites del Estándar CubeSat (Spanish Edition), 2021, ISBN-979-8510113068.

Made in the USA
Las Vegas, NV
10 February 2022

43627691R00028